MW00845927

A LOOK AT CHEMISTRY

ELEMENTS

BY KENNON O'MARA

Gareth Stevens
PUBLISHING

CRASHCOURSE

Please visit our website, www.garethstevens.com. For a free color catalog of all our high-quality books, call toll free 1-800-542-2595 or fax 1-877-542-2596.

Library of Congress Cataloging-in-Publication Data

Names: O'Mara, Kennon, author.
Title: Elements / Kennon O'Mara.
Description: New York : Gareth Stevens Publishing, [2019] | Series: A look at chemistry | Includes index.
Identifiers: LCCN 2018014327| ISBN 9781538230114 (library bound) | ISBN 9781538231418 (pbk.) | ISBN 9781538233283 (6 pack)
Subjects: LCSH: Chemical elements--Juvenile literature. | Chemistry--Juvenile literature.
Classification: LCC QD466 .O63 2019 | DDC 546--dc23
LC record available at https://lccn.loc.gov/2018014327

Published in 2019 by
Gareth Stevens Publishing
111 East 14th Street, Suite 349
New York, NY 10003

Designer: Reann Nye
Editor: Therese Shea

Photo credits: Series art Marina Sun/Shutterstock.com; cover Dorling Kindersley/Getty Images; p. 5 Richard Powers/ArcaidImages/Getty Images; p. 9 Phawat/Shutterstock.com; p. 11 Chaikom/Shutterstock.com; p. 13 MarcelClemens/Shutterstock.com; p. 17 (water) anmo/Shutterstock.com; p. 17 (Robert Boyle) Georgios Kollidas/Shutterstock.com; p. 19 (copper) Only Fabrizio/Shutterstock.com; p. 19 (iron) Kletr/Shutterstock.com; p. 19 (silver) joesfauer/Shutterstock.com; p. 19 (lead) Zelenskaya/Shutterstock.com; pp. 21, 23 Humdan/Shutterstock.com; p. 25 Vadim Sadovski/Shutterstock.com; p. 27 (pencil) studiovin/Shutterstock.com; p. 27 (diamond) Thomas Yeoh/Shutterstock.com; p. 29 (soda can) Scanrail1/Shutterstock.com; p. 29 (foil) Picsfive/Shutterstock.com; p. 29 (airplane) phive/Shutterstock.com.

Printed in the United States of America

CPSIA compliance information: Batch #CW19GS: For further information contact Gareth Stevens, New York, New York at 1-800-542-2595.

CONTENTS

Words in the glossary appear in **bold** type the first time they are used in the text.

IT'S ELEMENTARY!

Think of any object. Your games, toys, bed, even your brother—they're all made of elements! Elements are basic **substances**. They can't be broken down into simpler substances by any natural way. Elements make up all matter in the **universe**!

MAKE THE GRADE

About 90 elements exist in nature.

ABOUT ATOMS

Each element is made of one kind of atom. An atom is made of small **particles** called neutrons, protons, and electrons. If you could look into the atoms of an element, you'd see all the atoms have the same number of protons.

HELIUM ATOM

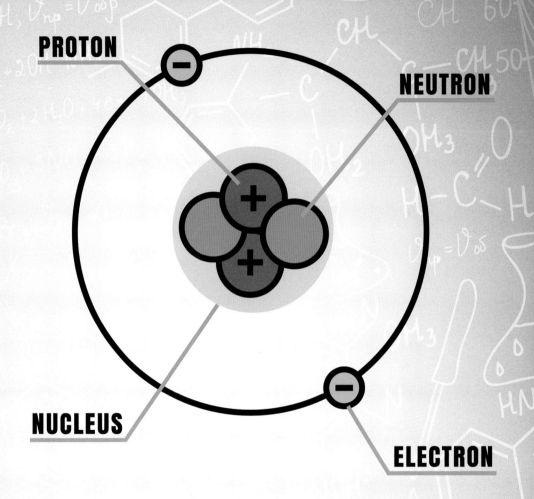

PROTON

NEUTRON

NUCLEUS

ELECTRON

MAKE THE GRADE

All helium atoms have two protons.

7

Atoms join with other
atoms to make molecules,
and molecules make the
substances around us.
Molecules of gold atoms
make up pure gold found in
the earth. Other elements,
such as hydrogen, are
hard to find in pure forms
in nature.

 MAKE THE GRADE

Some elements aren't found in nature, but have been created in labs by scientists!

9

Atoms of different elements can join to make molecules, too. These molecules are called compounds. Each molecule of the compound carbon dioxide has one atom of the element carbon and two atoms of the element oxygen. Plants use carbon dioxide to make food.

MAKE THE GRADE

Scientists have **identified** millions
of compounds so far!

SOLID, LIQUID, OR GAS?

Every element can be a solid, liquid, or gas. Very cold and very hot **temperatures** can change an element's form. That's because temperatures can affect how molecules act. Bonds between atoms can break or form, causing a change in matter.

MAKE THE GRADE

Mercury is the only common metal element that's liquid at room temperature.

MERCURY

The compound water is formed by the elements hydrogen and oxygen. These are both gases in their pure form. However, when two hydrogen atoms link to an oxygen atom, the features of the elements change. Together, these gases become a liquid.

THE COMPOUND WATER

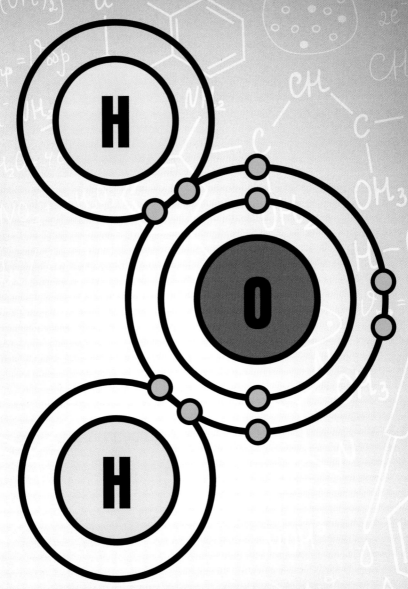

 **MAKE THE GRADE**

When water joins to the
compound sodium chloride,
salt water forms!

ELEMENTS OVER THE YEARS

Ancient Greeks believed there were just a few elements: water, fire, air, and earth. They thought these made up everything. In the 1600s, scientist Robert Boyle reasoned they couldn't be elements because they didn't form matter.

ROBERT BOYLE

MAKE THE GRADE

We know water isn't an element because it can be broken down into simpler substances. Elements can't be broken down.

Scientists began to **experiment** with identifying elements. As they became skilled in breaking down matter, elements were discovered. Scientists also began to understand more about the atoms that made up the elements. Later, they even saw atoms through **microscopes**!

COPPER

SILVER

LEAD

IRON

MAKE THE GRADE

The metals gold, silver, copper, iron, lead, tin, and mercury were among the earliest known elements because they're found in a pure form in nature.

ORGANIZING THE ELEMENTS

As scientists found out more about the elements, they realized the elements could be **organized** in a helpful way. This is how the periodic table came to be. Russian scientist Dmitri Mendeleev organized the first periodic table of elements in 1869.

Periodic Table of the Elements

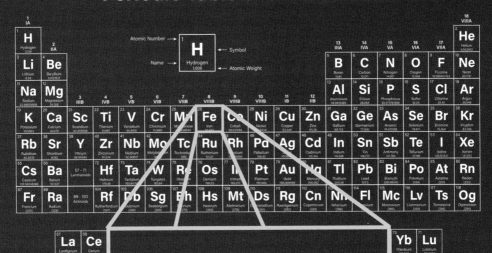

Fe
Iron

MAKE THE GRADE

On the table, each element's name is shortened to one or two letters. This is its atomic symbol. For example, "H" is for hydrogen. "Fe" stands for iron because the Latin word for "iron" is *ferrum*.

In the periodic table, elements are placed in rows according to the number of protons found in an atom of that element. This is the atomic number. Elements in a column share features with other elements in that column.

Periodic Table of the Elements

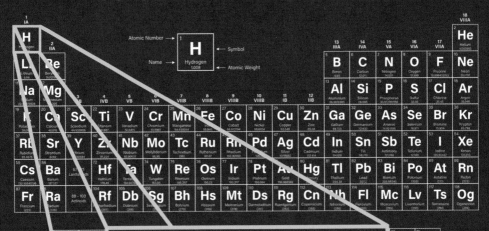

ATOMIC NUMBER

ATOMIC SYMBOL

ATOMIC WEIGHT

1

H

Hydrogen

1.008

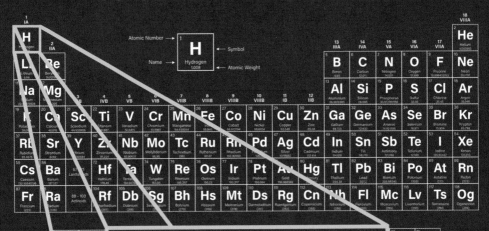

23

ELEMENTS AROUND YOU

The first 18 elements in the periodic table—the elements with atomic numbers 1 through 18—make up most of the matter in the universe. You've probably heard of these, but you might not identify them in everyday objects.

 MAKE THE GRADE

Elements on Earth are the same as elements on other planets!

25

Since nitrogen and oxygen are gases, you can't see them even though they're all around you. Depending how atoms of the element carbon bond, it can take different forms, called allotropes. Graphite and diamond are two allotropes of carbon.

GRAPHITE

DIAMOND

 MAKE THE GRADE

Graphite is used in some kinds of pencils.

27

Aluminum is an element used to make soda cans. It's also used to wrap food. When aluminum is combined with other metals, it becomes very strong. We use it to make airplanes. What would our world be without amazing elements like this?

ALUMINUM FOIL

SODA CAN

AIRPLANE

MAKE THE GRADE

Aluminum is the most **abundant** metal inside Earth, but it's only found in nature in compounds.

USEFUL ELEMENTS!

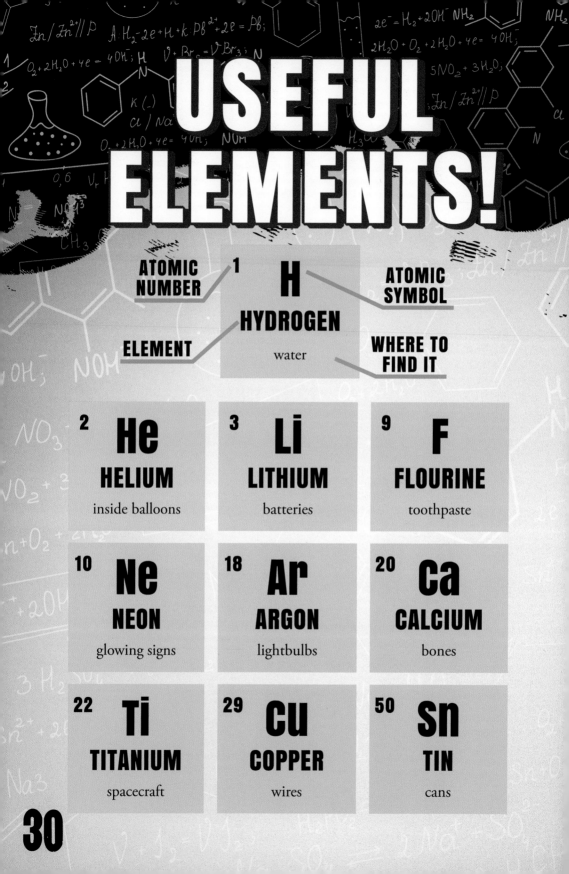

ATOMIC NUMBER 1

H

ATOMIC SYMBOL

HYDROGEN

ELEMENT

water

WHERE TO FIND IT

2 **He**
HELIUM
inside balloons

3 **Li**
LITHIUM
batteries

9 **F**
FLOURINE
toothpaste

10 **Ne**
NEON
glowing signs

18 **Ar**
ARGON
lightbulbs

20 **Ca**
CALCIUM
bones

22 **Ti**
TITANIUM
spacecraft

29 **Cu**
COPPER
wires

50 **Sn**
TIN
cans

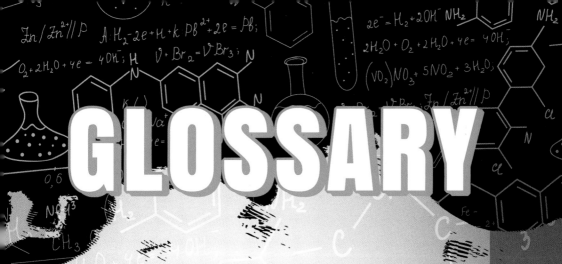

GLOSSARY

abundant: present in great amounts

experiment: to carry out a series of actions and watch what happens in order to learn about something

identify: to find out the name or features of something

microscope: a tool used to view very small objects so they can be seen much larger and more clearly

organize: to put together in an orderly way

particle: a very small piece of something

substance: a certain kind of matter

temperature: how hot or cold something is

universe: everything that exists

FOR MORE INFORMATION

BOOKS

Deschermeier, Charlotte. *Elements*. New York, NY: PowerKids Press, 2014.

Kukla, Lauren. *Elements at Work*. Minneapolis, MN: Sandcastle, 2017.

WEBSITES

Elements

www.ducksters.com/science/elements.php

Discover more facts about the elements that make up our universe.

Periodic Table and the Elements

www.chem4kids.com/files/elem_intro.html

Read more about some common elements as well as the periodic table.

INDEX